Earthquake Resistant Construction of Electric Transmission and Telecommunication Facilities Serving the Federal Government Report

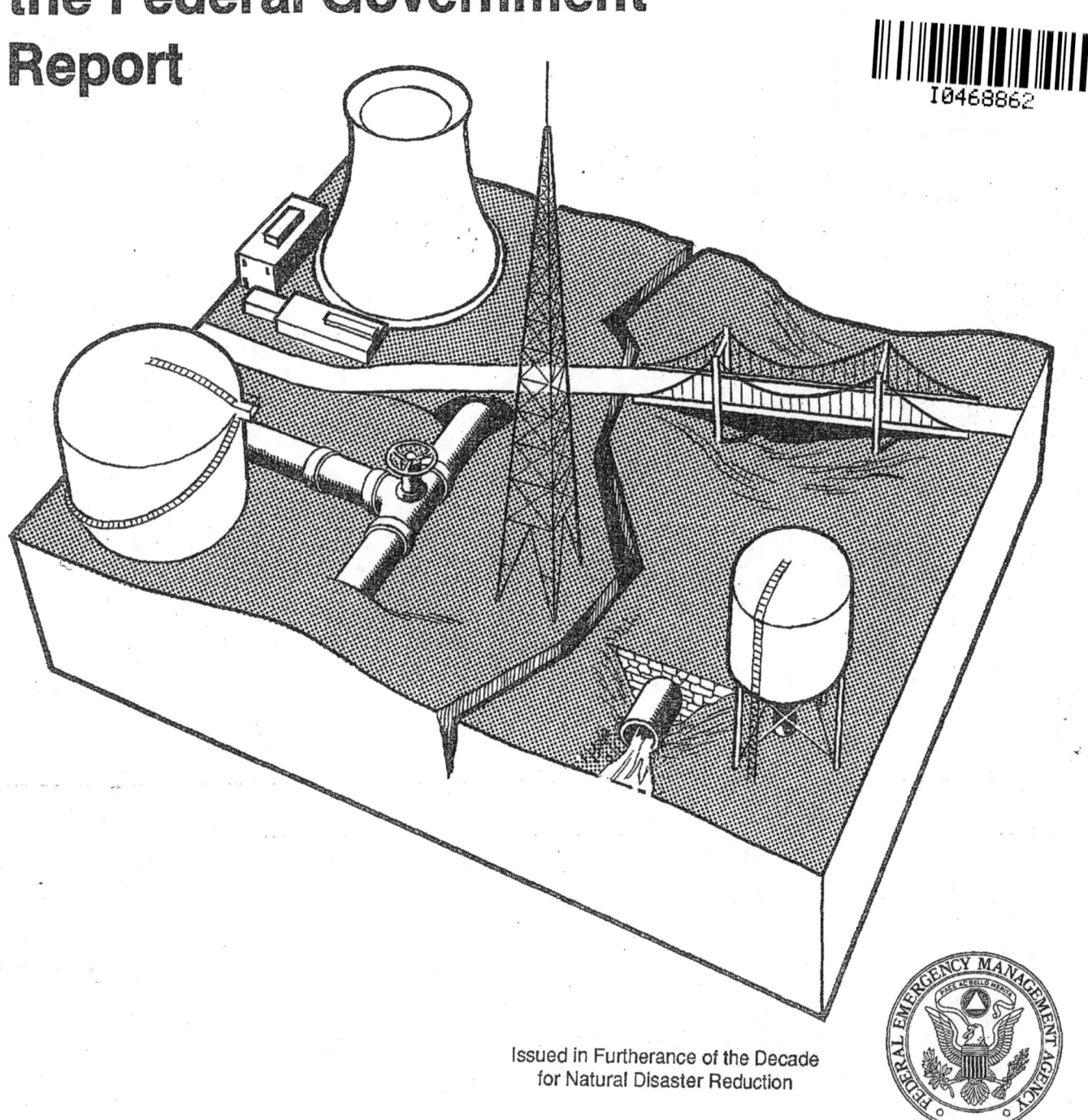

I0468862

Issued in Furtherance of the Decade
for Natural Disaster Reduction

Earthquake Hazard Reduction Series 56

NISTIR 89 - 4213

EARTHQUAKE RESISTANT CONSTRUCTION OF ELECTRICAL TRANSMISSION AND
TELECOMMUNICATION FACILITIES SERVING THE FEDERAL GOVERNMENT

Felix Y. Yokel

U.S. DEPARTMENT OF COMMERCE
National Institute of Standards and Technology
National Engineering Laboratory
Center for Building Technology
Gaithersburg, MD   20899

Prepared for

FEDERAL EMERGENCY MANAGEMENT AGENCY
500 C Street, S.W.
Washington, DC   20472

February 1990

U. S. Department of Commerce, Robert A. Mosbacher, Secretary

National Institute of Standards and Technology
John W. Lyons, Director

ABSTRACT

The vulnerability of electrical transmission and telecommunication facilities to damage in past earthquakes, as well as available standards and technologies to protect these facilities against earthquake damage are reviewed. An overview is presented of measures taken by various Federal agencies to protect electrical transmission and telecommunication facilities against earthquake hazards. It is concluded that while most new facilities which are owned and operated by Federal agencies are presently designed to provide some, though not necessarily adequate, earthquake resistance, there generally is no effort to retrofit existing facilities. No evidence was found of requirements to protect electrical transmission and communication facilities which have major contractual obligations to serve the Federal Government and only limited seismic design requirements are stipulated for electrical transmission systems constructed with Federal funding. It is recommended that Federal guidelines be developed for minimum levels of seismic design of electrical transmission and telecommunication systems.

Key words: central telephone offices; earthquake engineering; electrical power transmission; electrical substations; lifelines; seismic design standards; telecommunications

EXECUTIVE SUMMARY

The vulnerability of electrical transmission and telecommunication facilities to damage in past earthquakes, as well as available standards and technologies to protect these facilities against earthquake damage are reviewed. An overview is presented of measures taken by various Federal agencies to protect electrical transmission and telecommunication facilities against earthquake hazards.

It is concluded that, while most new facilities which are owned and operated by Federal agencies are presently designed to provide some, but not necessarily adequate, earthquake resistance, there generally is no effort to retrofit existing facilities. No evidence was found of requirements to protect electrical transmission and telecommunication facilities which have major contractual obligations to serve the Federal Government and only limited seismic design requirements are stipulated for electrical transmission systems constructed with Federal funding.

Records from past earthquakes indicate that electrical transmission lines are not very vulnerable to earthquake damage. However, earthquake damage occurred in areas of unstable soils. Substations were damaged in many earthquakes as a result of breakage of porcelain components, inadequate tiedown, inadequate slack in electrical lines, leaking gaskets, and inadequate clearance between system components. Distribution lines are not very vulnerable, however some damage occurred as a result of tangled wires and toppled platform-mounted transformers.

Central telecommunication offices and the equipment they house are in some instances vulnerable to earthquake damage. A large number of central telecommunication offices in the United States are located in seismically active areas. Telecommunication lines and microwave towers did not suffer much damage, except in areas of unstable soils.

There are also regional aspects of earthquake vulnerability. On the West Coast where there is a high frequency of seismic events, the need for earthquake resistant construction is recognized and electrical transmission and telecommunication systems are generally designed and built to be earthquake resistant. In seismically vulnerable areas on the East Coast and in the Central U.S. this is frequently not the case. The vulnerability of East Coast and Central U.S. systems is further increased because earthquakes in these regions tend to affect larger areas than West Coast earthquakes. Thus some of the more localized system redundancies which tend to protect electrical transmission and telecommunication lifelines on the West Coast may not function in a major earthquake in the Eastern or Central U.S.

There are some standards and guidelines for the earthquake resistant construction of electrical transmission lines and substation components, however there is no comprehensive document covering all aspects of the problem. ANSI/IEEE Standard 693-1984, 1984, is a recommended practice for seismic design of substations, but it is considered inadequate by industry and needs to be updated. There are also criteria in the model building codes [Uniform Building Code, 1988, BOCA Building Code, 1986, Southern Building Code, 1989], the SEAOC Tentative Lateral Force Requirements, 1985, and the NEHRP Recommended Provisions, 1989, which could be used to design foundation tiedowns and components of electrical substations, however the provisions in these documents are primarily used in conjunction with building codes which do not necessarily apply to the design of electrical substations. Several design manuals published by military agencies also contain provision that are applicable to electrical substations. Utility companies on the West Coast substitute their own specification for ANSI/IEEE Standard 693, to insure adequate earthquake resistance.

Technology is available, and presently used in new construction on the West Coast, which can provide adequate components, housing and tiedown for substation equipment. Adequate clearances and slack in connecting lines can also be provided, even though some configurations which provide protection against earthquake loads are not always the most efficient from the standpoint of electrical performance. However, present technology does not provide adequate protection of, or a satisfactory substitute for, porcelain components. Thus even substations which are designed in accordance with the best available technology are susceptible to some earthquake damage. Efforts are now under way to develop materials which can replace presently used vulnerable porcelain components.

Standards for telecommunication centers and generic technical requirements for new equipment are being developed by Bellcore and some other industry organizations. Standards developed for central office equipment and its tiedown and bracing systems are designed to provide adequate protection. These standards are being updated on an ongoing basis. Bellcore also developed operational procedures for earthquake emergencies and hazard mitigation. Compliance with these industry standards is voluntary. The standards are followed closely on the West Coast, where existing facilities are also often retrofitted to comply with the standards. In other parts of the country the emphasis on earthquake protection is generally less and varies among regions. Central offices are presently designed in compliance with local building codes which call for earthquake resistant design. However, attention should be given to the assignment of appropriate importance factors (seismic hazard exposure) to these buildings. There is a trend for increased concentration and loss of redundancy of communication facilities which is brought about by the high capacity of optical fiber cables. As a result, the potential disruptions associated with failures of communication centers could be more serious.

There is presently no indication that facilities designed in accordance with Bellcore recommendations, using adequate design ground accelerations, would be vulnerable to earthquake damage. However, it is important to note that as a result of divestiture, equipment is now being purchased from a variety of U.S. and foreign sources. Unless seismic requirements are incorporated in the specifications for this equipment, it may be vulnerable to earthquake damage.

There are three types of lifelines which could be required to meet Federally-imposed standards: lifelines which are owned and operated by the Federal Government; lifelines which are owned and operated by others, but have a major contractual obligation to serve the Federal Government; and lifelines which are constructed with Federal funds but not owned by the Federal Government and which do not have a major contractual obligation to serve the Federal Government.

Most new construction projects for electrical transmission systems which are owned and operated by the Federal Government are designed to be earthquake resistant. However, their design and construction specifications are generally not as stringent as those followed by private utilities on the West coast. Also, different agencies follow different rules. At the present time not much retrofitting of inadequate existing facilities is planned.

There is no evidence that Federal Agencies require compliance with the Bellcore recommendations or comparable standards by other carriers when telecommunication equipment for seismically vulnerable areas is purchased, or that G.S.A. schedules for the purchase of this equipment incorporate earthquake-resistant components. There is also no evidence that privately owned and operated electrical transmission and telecommunication facilities which have a major contractual obligation to the Federal Government are required to be protected against earthquake hazards. However, some of the operators of these facilities may provide such protection on their own initiative.

Electrical transmission facilities constructed with Federal funds and to Federal specifications, but not serving the Federal Government, are required to provide some earthquake protection; but the scope of these requirements is limited and does not include references to the presently-used standards and practices discussed in this report.

It is recommended that inclusive Federal guidelines for minimum levels of seismic design of electrical transmission and telecommunication systems be developed under the auspices of the Interagency Committee for Seismic Safety in Construction. Since many Federal Agencies are opposed to a Federally imposed uniform standard, it is recommended that the guideline consist of two parts: (1) performance criteria; and (2) a model standard document.

The performance criteria should specify required levels of performance, without specifically stating how that performance is to be achieved. This will permit various agencies to make their own choice whether to achieve the required performance by following a specific standard or by developing their own provisions. The model standard is envisioned as a compilation and updating of existing standards.

As a first step in the development of this document, joint Federal-industry workshops should be organized, in order to assess the adequacy of existing standards and guidelines and identify the areas that need more work.

TABLE OF CONTENTS

ABSTRACT   iii

EXECUTIVE SUMMARY   v

1. INTRODUCTION   1

2. ELECTRICAL TRANSMISSION AND DISTRIBUTION SYSTEMS   2

    2.1 System Elements and Components   2
         2.1.1 Subsystems   2
         2.1.2 Substations   2
         2.1.3 Transmission lines   4
         2.1.4 Distribution systems   4
         2.1.5 Control and backup facilities   4

    2.2. Performance Record in Earthquakes   4

    2.3 Available Standards, Design Guides and Remedial Measures . . .   7
         2.3.1 Standards and Design Guides   7
         2.3.2 Remedial Measures   11
         2.3.3 Retrofit Versus Gradual Replacement   11

3. TELECOMMUNICATION SYSTEMS   12

    3.1 System Elements and Components   12
         3.1.1 Subsystems   12
         3.1.2 Distribution Facilities   13
         3.1.3 Telecommunication Buildings   14

    3.2 Performance Record in Earthquakes   17

    3.3 Available Standards, design guides and Remedial Measures . . .   18
         3.3.1 Standards and Specifications   18
         3.3.2 Remedial Measures   19

4. PROTECTION OF FEDERALLY-CONTROLLED SYSTEMS   21

    4.1 Introduction   21

    4.2 Hazard-mitigation Measures Implemented by Federal Agencies . .   22

5. SUMMARY AND RECOMMENDATIONS   27

    5.1 System Vulnerability   27

    5.2 Existing Standards, Guidelines, and Protective Measures . . .   28

    5.3 Federal Practices   29

    5.4 Recommendation   31

6. ACKNOWLEDGEMENT   31

7. REFERENCES   33

APPENDIX   37

1. INTRODUCTION

Within the framework of the Action Plan for the Abatement of Seismic Hazards to Lifelines (Building Seismic Safety Council, 1987), the National Institute of Standards and Technology (NIST) reviewed measures presently taken by Federal Agencies to protect electrical power transmission and telecommunication lifelines against seismic hazards. This report summarizes the result of the study.

While the Federal Government extensively utilizes power transmission and communication facilities, only a small portion of these facilities are Federally operated. Thus, it is important to realize that even if all Federally-operated facilities were adequately protected against seismic risk, a large portion of the systems on which the Federal Government depends may still be vulnerable, unless measures to mitigate seismic risks are also implemented by private and regional utility companies.

A potential obstacle to the implementation of all the protective measures which are in the public interest is that the cost-effectiveness of mitigating seismic risks for most utilities must be calculated in terms of lost equipment and revenues. This does not necessarily reflect the risk to the public at large or to government agencies associated with the potential for loss of life and health and lost production. Another problem is that electrical power systems, and to a a lesser extent communication systems, may not be very vulnerable to moderate earthquakes because of system redundancies, but they could be wiped out by a great earthquake. However, in most parts of the U.S., the recurrence interval of great earthquakes is such that adequate hazard mitigation against such events is not always considered cost-effective by utility companies and the public service commissions which regulate their rates.

In some instances engineers responsible for the design of Federal facilities use the state of the art rather than specific agency-promulgated standards or criteria to design system components. Thus, an effort is made to design new Federal facilities against earthquake hazards, even if the standards of the Agency do not explicitly require earthquake-resistant design.

There are presently miscellaneous sources of information which can be used in the design of earthquake-resistant facilities, however there is no single design guide which deals with all aspects of the problem. While much has been learned from past earthquakes, recent earthquake records show that electrical facilities which have been designed to be earthquake resistant can be severely damaged in a relatively moderate earthquake EPRI NP-5607, 1988. Thus, there is a need to improve design and construction techniques.

Sections 2 and 3 of the report deal with the seismic vulnerability of electrical transmission and telecommunication facilities, respectively, and with presently-used standards, design criteria and earthquake-resistant components and systems; Section 4 summarizes Federal practices in the design of new facilities and the retrofit of old facilities; and conclusions and recommendations are presented in Section 5. Statements made by engineers from various Federal agencies and other organizations contacted in this study are summarized in the Appendix.

2. ELECTRICAL TRANSMISSION AND DISTRIBUTION SYSTEMS

2.1 System Elements and Components

2.1.1 Subsystems

Electrical power transmission systems can be divided into <u>four</u> subsystems: **substations**, which according to past experience are vulnerable to seismic damage; **transmission lines**, which in the past have not been very vulnerable to earthquake damage; **distribution systems** which are also not very vulnerable to earthquakes; and **control and backup facilities** which are occasionally damaged because of inadequate tiedown and support systems.

2.1.2 Substations

Substations fall into several categories: <u>Transmission substations</u> reduce transmission voltage to subtransmission voltage for distribution to distribution substations. They have large transformers and switching and controlling equipment. <u>Switching substations</u> (switchyards) provide circuit protection and system switching flexibility. They have switching and controlling equipment. Important substation components are listed below:

Structures and foundations: Substations contain structures as well as equipment mounted outside on footings. Anchorage to footings, supporting equipment, is frequently inadequate to prevent overturning or slipping in earthquakes. Tiedown to foundations is of particular concern in earthquake resistant design.

Power Transformers transform voltage in the main power system. They are equipped with a pressurized oil cooling systems contained in a tank surrounding the transformer which is either pressurized by an inert gas layer or through a connection with a second tank (conservator). Transformers can be self-cooled, use a forced-air cooling system or use a forced-air and a forced-oil cooling system. The transformers themselves, their foundation connections and cooling systems, bushings, as well as other equipment mounted on them or connected to them by wires can be vulnerable to seismic loads.

Power circuit breakers are devices that close and open electrical circuits between separable contacts under load and fault conditions. The separating medium can be oil, air, gas or a vacuum. Circuit breakers are either "dead tank" (maintained at ground potential) or "live tank" (maintained at line potential and mounted high on an insulating porcelain column). Operating mechanisms can be solenoids, motors, pneumohydraulic or pneumatic devices, motor or manually charged springs, or manual.

Other major equipment includes reactors, capacitors, air switches, wave traps, surge arresters, disconnect switches, current transformers, capacitive voltage transformers and lightening arresters.

Other items include equipment support structures, miscellaneous transmission lines and raceways, busses and miscellaneous porcelain columns, bushings and insulators, standby power supplies (batteries, generators or circuits), control panels and equipment, and water and oil storage tanks.

2.1.3 Transmission lines

Transmission lines include towers and poles and their foundations, insulators, conductors, and ground wires and underground cables with their potheads and oil circulating and cooling equipment.

2.1.4 Distribution systems

Distribution systems include poles, pole-mounted transformers, and above-ground, as well as underground conductors.

2.1.5 Control and backup facilities

These facilities include computer-assisted control centers, repair and maintenance facilities, and communication systems.

2.2 Performance Record in Earthquakes

The performance of electrical transmission and distribution systems in earthquakes that occurred before 1981 is summarized by ASCE, 1983, 1987[1]. More recently, the Electric Power Research Institute (EPRI) prepared studies of five recent earthquakes with special emphasis on mechanical and electrical control equipment EPRI NP-5607, 1988, NP-4605, 1986, NP-5616, 1988, NP-5784, 1988, NP-5970, 1988. These studies contain important information on components which are part of the electrical transmission system. Information available from the 1989 Loma Prieta earthquake (Benuska, 1990) is also included in this report.

Table 1 (p.5) contains a partial summary of accumulated information. It can be seen from the tabulation that transmission towers and poles are not very vulnerable. However, damage was reported in the 1923 Kanto earthquake and in the 1964 Alaska earthquake. The reported damage in the Kanto earthquake was attributable to land slides and that experienced by the Chugak Power company in

---

[1] References are identified by author and year of publication and listed alphabetically in Section 6.

4

Earthquakes:

No.	Earthquake	Year	M*	PGA**
(1)	Kanto, Japan	1923		
(2)	Kern County, CA	1952	7.7	0.20-0.25
(3)	Alaska	1964	8.4	0.10-0.30
(4)	San Fernando, CA	1971	6.5	0.20-0.60
(5)	Managua, Nicaragua	1972	6.2	0.40-0.60
(6)	Miyagi-ken-oki, Japan	1978	7.4	0.10-0.40
(7)	Imperial Valley, CA	1979	6.6	0.50
(8)	Corinth, Greece	1981	6.7	0.15-0.30
(9)	Palm Springs, CA	1986	5.8	0.72-0.97
(10)	San Salvador	1986	5.4	0.50
(11)	New Zealand	1987	6.3	0.3(-1.0?)
(12)	Loma Prieta, CA	1989	7.1	0.64

Damage Matrix:

System Component \ Earthquake No.	1	2	3	4	5	6	7	8	9	10	11	12
Porcelain Insulators Broken			x	x	x	x	x	x	x	x	x	x
Transformers Tipped Over and damaged			x	x						x	x	x
Switchgear & Cabinets shifted/overt.			x	x	x					x#	x#	x
Circuit Breakers Damaged			x	x					x		x	x
Transformer Conductors Broken			x	x	x							x
Buckled Conduit			x									x
Pole Transformers Overturned		x										x
Distribution Wires Entangled		x		x	x							x
Transmission Towers Destroyed/Damaged	x	x										
Excessive Relative Displacements	x	x	x	x			x			x	x	x

* Richter Magnitude
** Peak Ground Acceleration
# Damage Caused By Inadequate or Lacking Anchorage

5

Alaska was confined to lines in areas of alluvial deposits ASCE, 1987. Transmission towers in the same general area which were located on rock were not damaged. This low vulnerability of towers is probably related to the fact that the natural frequency of most towers in the horizontal vibration mode is low, and therefore the wind forces on the tower and the supported wires, which transmission towers are designed to resist exceed the forces generated by earthquakes. Thus failures that do occur are mostly related to foundation instability. However, damage has been reported for a microwave tower and for hilltop antennas from the Loma Prieta earthquake.

Failure of porcelain components such as insulators, bushings, and support columns is very common. This is probably the greatest problem facing electrical utilities in earthquake areas, because substitute materials have not been sufficiently evaluated to warrant their large-scale introduction. Some of the porcelain damage, and also damage to other equipment components and busses was caused by inadequate slack between elements connected by wires, or by bumping because of inadequate clearance. Typically, many of the failures of major equipment components such as transformers and circuit breakers are caused by breakage of porcelain components such as bushings and support columns EPRI NP-5607, 1988. High-voltage equipment is generally more vulnerable than lower-voltage equipment because of greater clearance requirements, which in turn require taller ceramic columns.

Transformers and their conductors were damaged in many earthquakes because of inadequate tiedown. However not all transformers which experienced foundation tiedown failures were damaged. Leaking gaskets are another frequently observed form of transformer damage. The resulting loss of cooling fluid in turn causes the transformer to fail. Damage to power transformers is very disruptive and may take a long time to repair. Circuit breakers were damaged in several earthquakes. The live-tank type circuit breakers[2] which have interrupter heads mounted on porcelain columns are particularly vulnerable. Air switches also are

---

[2] The housing containing the interrupting mechanism of "live-tank" circuit breakers is maintained at line potential, and therefore must be mounted on an insulating porcelain column. The circuit-breaker tank and all accessories of "dead-tank" circuit breakers are maintained at ground potential.

vulnerable to damage, as are lightning arresters which are mounted on porcelain columns.

Many other types of major equipment also suffered damage by breakage of porcelain, inadequate tiedown, and excessive relative motion between components. Damage also occurs when minor equipment and accessories, such as batteries, battery racks, and electrical panels are unsecured or inadequately secured.

Information from the 1989 Loma Prieta, CA earthquake (Benuska, 1990) also demonstrates the vulnerability of substations, and particularly of substation elements with porcelain components. Four substations suffered damage. Three of these substations were severely damaged. Particularly vulnerable were ceramic columns supporting circuit breakers, bus-support structures, disconnect switches, and bushings and radiators of transformers. Service was restored by temporarily bypassing some of the damaged circuit breakers.

Distribution systems are generally more rugged than high voltage transmission systems and the poles and lines are not very susceptible to damage. However, damage can occur by failure of adjacent structures, and distribution wires may become entangled and burn, often before the circuit breakers respond and cut off the current. There was also extensive damage to platform mounted transformers which toppled over in the 1952 Kern County earthquake. Some platform-mounted transformers also toppled in the Loma Prieta earthquake.

2.3 Available Standards, design guides and Remedial Measures

2.3.1 Standards, Codes, and Design Guides

Several standards and specifications for earthquake resistant design are available at the present time. Structures and foundations are adequately covered in present and proposed building codes, standards and resource documents, such as the Uniform Building Code, 1988, the BOCA National Building Code, 1986, the Standard Building Code, 1989, the National Earthquake Hazards Reduction Program (NEHRP), "Recommended Provisions for the Development of Seismic Regulations for New Buildings", 1988, and the Structural Engineers

Association of California (SEAOC) "Tentative Lateral Force Requirements", 1985. These may have to be supplemented by some additional information when there is a need to control stiffness as well as strength of structures, possibly via assignment of appropriate importance factors (seismic hazard exposure) to these buildings. The above-mentioned standards and resource documents also contain mechanical and electrical design requirements. Even though these requirements are primarily addressed to building design, they are generally written broadly enough to also be applicable to lifeline systems. For example Section 8 of the NEHRP Recommended Provisions addresses mechanical and electrical systems. The provisions stipulate lateral force requirements for the design of attachments and require a manufacturer's certification for important components, which specifies that the components shall not sustain damage under specified equivalent static forces. While such provisions could be applied to the design of electrical transmission system components, their adoption by local building codes would not insure their enforcement for lifelines which are not under the jurisdiction of building inspectors.

Seismic design of electrical equipment in substations is covered by the American National Standards Institute (ANSI) and the Institute of Electrical and Electronics Engineers, Inc.(IEEE), ANSI/IEEE Standard 693-1984, which in turn references ANSI/IEEE Standard 344-1987, "IEEE Recommended Practice for Seismic Qualification of Class 1E Equipment for Nuclear Power Generating Stations". The standard identifies Class A equipment or systems, whose failure would prevent the proper functioning of the substation, and equipment of lesser importance. Equipment can be qualified by testing, analysis, or on the basis of prior experience. The standard contains performance criteria for all substation components. However, its most important features are criteria for the qualification of equipment. These criteria, together with a specified maximum design ground acceleration, specify the seismic resistance of equipment to be furnished by suppliers. West Coast power companies consider the present ANSI/IEEE Standard 693 inadequate and substitute their own specifications for the performance of individual components. Figure 1 shows design spectra specified by the City of Los Angeles for various damping ratios for a maximum ground acceleration is 0.5 G. The accelerations in the spectra are scaled up or down in accordance with the stipulated maximum ground acceleration.

8

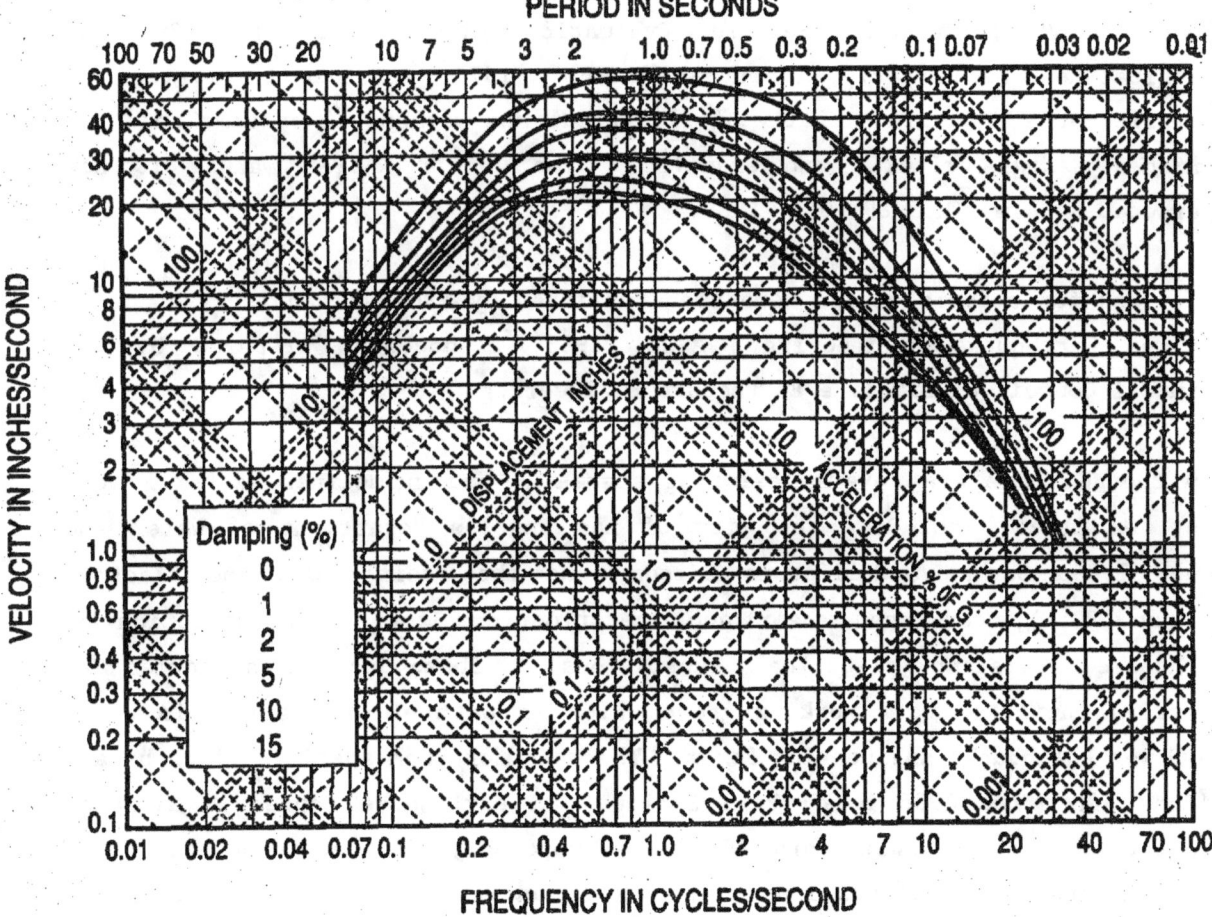

Figure 1:   Seismic Design Response Spectra for Electrical Substation Equipment
(from City of Los Angeles, 1988).

Seismic standards and technical manuals are also published by military agencies. These include the Tri-Service manuals "Seismic Design for Buildings",1982, and "Seismic Design Guidelines for Essential Buildings", 1986, which contain amendments to the SEAOC Tentative Lateral Force Requirements, 1985, which are specifically applicable to flexible or flexibly mounted mechanical and electrical equipment. Some seismic requirements are also contained in the Military Handbook "Electrical Utilization Systems", 1987, but these are limited to recommended choices of systems which are less vulnerable, such as overhead wires in preference to underground cables. In addition, the U.S. Army Corps of Engineers developed a guide specification, 1985, and the Naval Facilities Command (NAVFAC) manual "Seismic Evaluation of Supports for Existing Electrical-Mechanical Equipment and Utilities", 1975, provides information on the performance and vulnerability of electrical and mechanical systems and also furnishes design criteria and examples. The latter NAVFAC manual is now superseded by Army TM 5-89-10-1, 1986.

Guidelines for seismic design were also published by ASCE as "Advisory Notes on Lifeline Earthquake Engineering", ASCE, 1983, a document prepared by the Technical Council on Lifeline Earthquake Engineering.

Even though there are several existing standards and guidelines, several West Coast utilities, including the City of Los Angeles Department of Water and Power, 1988, Southern California Edison, and the Pacific Gas and Electric Company (PG&E), developed their own specifications for the construction of earthquake resistant electrical substations. These companies formed an electrical utility staff working group which is part of the California Water and Power Earthquake Engineering Forum. The need for such specifications became evident after the 1986 North Palm Springs, CA earthquake caused significant damage to the Devers switchyard, which was supposedly designed to be earthquake resistant (based on lessons learned from the San Fernando earthquake, EPRI NP-5607, 1988, Schiff, 1973). Even though ground accelerations exceeded those anticipated in the design by almost 100%, the failure gave an indication that design practices need to be improved. Part of the problem is the vulnerability of porcelain components. However there are other problems, including the need to reconcile efficient electrical design with requirements for seismic ruggedness. The two concepts are often conflicting. The Electrical Power Research Institute (EPRI) is also currently developing material which will help improve seismic performance od substations in cooperation with the West Coast power companies.

2.3.2 Remedial Measures

Several remedial measures against earthquake damage are presently available. Buildings and foundations for substations and their components, located in seismic areas, can be designed by the applicable building codes. This procedure, however, is seldom implemented for transmission tower foundations which are not regulated by the building codes. Substation components in vulnerable areas can be tied to their foundations using existing criteria. Allowance can be made for earthquake-induced relative motions between components. IEEE procedures for qualifying equipment can be specified when ordering equipment (the acceleration level must be specified). However there are some supply problems because of limited demand. Some of the available earthquake resistant hardware components are listed by Steinhardt, 1987.

The Electrical Power Research Institute (EPRI) developed Polysil, a polymer-silicone based material which shows promise as a substitute for porcelain (refer to "Filled Polymer"); however, there are still some questions about long term creep under load and about the economics of the material. The specific purpose of developing this latter material was not necessarily seismic resistance. Stronger porcelains are also being developed (refer to "High Strength Alumina"). More recently, a fiber-glass/silicon rubber mixture has been used as a substitute for porcelain bushings. Some other techniques can also improve seismic performance, such as shock-resistant mountings, switching from the porcelain-mounted live tank circuit breakers to dead tank circuit breakers, and generally replacing many vulnerable components by more rugged hardware.

2.3.3 Retrofit Versus Gradual Replacement

The point has been made that seismic design provisions should primarily

11

concentrate on new construction and not emphasize retrofit (ASCE, 1983). In many instances this makes sense, because in many parts of the United States the life cycle of much of the equipment is shorter than the recurrence interval between major seismic events, and chances are that when the next major earthquake hits much of the current equipment will have been replaced. However some equipment types, such as transformers, have a long service life, and retrofit measures to protect them, such as adequate tiedown, are relatively inexpensive. Thus at least some retrofit measures should receive consideration, as they do in California. The cost effectiveness of these measures to the public is much greater than that calculated on the basis of replacement cost and lost revenue for the utility company, because of the potential losses in service and production associated with a disruption of the power supply.

3. TELECOMMUNICATION SYSTEMS

3.1 System Elements and Components

3.1.1 Subsystems

It is convenient to identify the elements of the physical constituents of telecommunication systems as parts of two subsystems (Foss, 1981): (1) **distribution facilities** which include elements that transmit the messages; and (2) **telecommunication buildings** which house switching equipment and terminal facilities for the transmission system. In general, distribution facilities are not very vulnerable to earthquakes, but buildings and the equipment they house are vulnerable. One of the most spectacular failures was the collapse of the SCT (Secretaria de Communicaciones y Telegrafos) communication center in the September 1985 Mexico City earthquake (Stone et al., 1987), which completely disrupted telephone communications to other countries for an extended period of time; and, as a consequence, impeded the coordination of rescue efforts.

3.1.2 Distribution Facilities

Distribution facilities consist of: (1) loop systems; (2) exchange area systems; and (3) long-haul systems.

(1) Loop systems

Loop systems leave central offices in underground conduits which are connected to smaller buried feeder cables. The feeder cables, in turn, are connected through a feeder/distribution interface to distribution cables which contain many pairs of loops for individual customers. The loops convey acoustic-pressure generated electrical signals to the central office. Distribution cables can be buried (not in conduits) or suspended on poles. Those suspended on poles derive their structural tensile strength from steel wires in their center. These systems, also referred to as "outside plant facilities" (located outside the central office buildings) also include poles and miscellaneous accessories, concrete manholes, and carrier equipment installed on poles or in concrete manholes. No seismic design requirements are used for these systems except for slack provided in underground cables crossing known faults (Faynsod, 1987). However, they are designed for wind and thermal loads, road vibrations, and transportation shocks. These latter loads generally exceed those imposed by seismic events.

(2) Exchange area systems

Exchange area systems contain many relatively short (typically a few miles long) trunks. Short trunks carry voice-frequency transmissions (4 kHz range) in a two-wire mode. Longer trunks use primarily digital carrier systems in four-wire circuits with a higher frequency range. These high-frequency lines are supported by terminal equipment for modulating and multiplexing (combining voice channels). There are also some high-capacity transmission lines for shorter hauls using coaxial cables or radio transmission.

Underground support facilities include reinforced concrete or masonry manholes, telephone exchange cable vaults, and concrete (multi-cable) or plastic single-cable) conduits between manholes and from exchange buildings to manholes, constructed in utility rights-of-way according to established utility standards. These facilities are not regulated by local building codes (Faynsod, 1987). However, they are designed to requirements which are in many instances more stringent than those needed for seismic resistance, in order to provide mechanical integrity and water tightness and to support loads imposed by automotive traffic.

(3) Long-haul systems

Long-haul systems provide long circuits and are designed to provide good performance for distances of several thousand miles. They use coaxial and fiber optics cables, submarine cables, microwave radio relay, and satellite radio. Microwave systems consist of microwave dishes or horns mounted on poles, buildings, or steel towers. The transmission equipment is mounted under the towers or poles. The towers are designed to resist wind forces which exceed anticipated seismic forces. Fiber optics is presently used for most long distance telecommunication transmissions. However, microwave installations are now extensively used in conjunction with cellular phones. As previously noted, a microwave tower suffered damage in the 1989 Loma Prieta earthquake. Microwave dishes mounted on buildings could also be vulnerable to earthquake damage.

3.1.3 Telecommunication Buildings

(1) Network of Central Offices

For long-distance calls, central offices are arranged in a hierarchical network (Foss, 1981). Calls first reach a local area central dial office, which in rural areas is typically a 1- to 3-story building. In urban areas, taller buildings are used for the most part. From the local area central dial office, the traffic may move through several toll centers which are larger structures. Finally there are primary centers which, generally, are tall buildings. It has been estimated (Shinozuka, 1987) that more than 15,000 United States communication centers are exposed to some seismic risk. Some of these buildings may not meet present standards for earthquake resistant design.

14

(2) Central office buildings

Buildings for communication centers, designed for equipment used in the past, have typically 12 ft-6 in. to 17 ft floor-to-ceiling heights (higher for older equipment) and 20 ft column spacings, and are designed for larger floor loading than ordinary office buildings (150 lb/ft^2 for equipment and overhead cable distribution systems alone). Since much of the floor load is due to equipment, no load reductions similar to those permitted in ANSI A-58.1, 1982 were used in their design. Since the floor loads in these buildings are large and there are relatively few outside windows, it is reasonable to assume that if the buildings were designed for seismic loading, they probably are stiffer than ordinary office buildings even if no special design provisions are applied. However, many of the older central office buildings are unreinforced masonry structures which are very vulnerable to earthquakes (Foss, 1981). Even though in present practice there is no explicit requirement to design central office buildings to meet strength, as well as stiffness criteria more stringent than those used in the applicable building codes, it has been suggested (Mirzad, 1987) that these buildings should be stiff, symmetrical shear-wall type structures. One way to insure better performance would be via the "importance factor" (UBC, 1988) or the "Seismic Hazards Exposure Group" (NEHRP, 1989).

New products, which are now being developed for a reduced equipment height of 6 ft, will require less ceiling height than the old equipment. New equipment technologies have made present electronic equipment more compact and lighter in weight than the older equipment, which used mechanical relay technology. Thus, as a consequence of equipment replacement, many buildings do not have the equipment densities and floor loads they once had and there are vacant spaces. As a consequence, the earthquake forces acting on these buildings are reduced.

(3) Central Office Equipment

Central office equipment includes switching equipment, cable handling systems, battery power supplies and other standby power sources, computer facilities and air conditioning systems.

Switching equipment: The older, 11-ft tall electro-mechanical switching equipment which was connected at the ceiling level is now being replaced with electronic equipment which is 6 ft tall and therefore does not lend itself readily to overhead bracing (Tang, 1984). Instead it is tied to the floor, acting like an inverted pendulum. This is not a disadvantage, since recent industry tests and studies indicate, that equipment which is designed to be independent of overhead bracing is less vulnerable to earthquake damage. Overhead bracing introduces out-of-phase motions between the ground support and the overhead auxiliary framing, resulting in drag forces on the equipment. The electronic equipment uses printed circuit boards which are sensitive to vibrations and could crack and can also be damaged by differential movements between the boards and the frame to which they are connected. Older equipment is vulnerable to earthquake forces which can cause disruption by the loosening of connections. In accordance with information obtained from Pacific Bell shake table tests of modern electronic equipment indicate that the equipment can tolerate considerable shock and vibrations.

Cable handling systems consist of a grid of framing suspended from the ceiling by hanger rods. Cable ladders and racks are mounted to the framing. As central offices expand, cable loads increase, requiring larger hanger-rod capacities.

Battery racks and the batteries they support are very vulnerable unless they are secured. When secured, racks are usually braced in the transverse direction between the top of the battery stands and the ceiling and by pairs of longitudinal braces at the end of each battery stand.

Computer equipment related to communications is normally installed on raised floors. Earthquake resistant floor systems have been developed, and should be used where appeopriate.

Air conditioning equipment is vital for the long term reliability of the system. However, equipment is designed to function up to two weeks in a 120° F environment. Requirements of local building codes for securing air conditioning equipment and the water and electricity supply needed for its proper functioning are not necessarily adequate for communication centers.

16

3.2 Performance Record in Earthquakes

Many earthquakes caused damage to specific telecommunication equipment. These include the 1933 Long Beach, CA earthquake which caused Pacific Bell to introduce seismically braced equipment (Foss, 1987), the 1971 San Fernando earthquake which caused severe damage to the Sylmar central office, the 1978 Myagi-Ken Oki, Japan earthquake which among other things damaged intercity coaxial cables in many locations (Shinozuka, 1987, Yanev, 1984), and the 1985 Mexico City earthquake where a major communication center was destroyed (Stone et al., 1987).

The effects of the 1971 San Fernando earthquake and various other earthquakes are summarized by Isenberg, 1984. In the Sylmar central office there was damage to switching equipment. There were also instances of damage to transmission lines (mostly to poles in areas of poor soil conditions), batteries and generators. There was one instance of damage to microwave equipment (Managua, Nicaragua) mounted on the fourth floor of a building. Shinozuka, 1987 summarizes the damage caused in previous earthquakes as: damage to equipment racks; bent or buckled frames; batteries jolted out of place; toppled transmission poles; and broken cables. Added to this list should be secondary effects caused by toppling of structures or failure of structural or non structural elements (such as suspended ceilings) which in turn destroy or damage communication equipment. Also failures in the water and power supply may disrupt the operation of essential air conditioning equipment.

The most severe problem is associated with central office equipment. This equipment is often difficult to replace. For instance in the San Fernando earthquake there was $4.5 million damage to switching equipment; and it took about 4 months to restore service at the Sylmar facility (Shinozuka, 1987). In accordance with information obtained from Pacific Bell, the turn-around time for new equipment has been demonstrated to be shorter in the repair of fire damage which occurred in 1988 at the Hinsdale, IL office. Limited service can be restored within several days, but restoration of the full capability would take longer.

17

In the 1989 Loma Prieta earthquake, telecommunication lifelines performed well. Central offices in Watsonville and Santa Cruz, which were close to the epicenter, continued to function and did not suffer significant damage. In the Watsonville communication center, equipment, which was subjected to maximum horizontal accelerations of 0.67 G and maximum vertical accelerations of 0.66 G, did not suffer damage and continued to function. The only service problems that arose in the Loma Prieta earthquake were power interruptions, caused by malfunction of stand-by generators which was not related to earthquake damage, dislocation of computer equipment in the Oakland central office, which suffered some earthquake damage, and dislocation of some brackets supporting cable trays (Benuska, 1990). It is also noteworthy that telecommunication manholes located in areas where soil liquefaction occurred did not suffer damage [in some previous earthquake manholes in liquefaction areas were damaged by uplift (buoyancy) forces]. Also, underground telephone cables in liquefaction areas did not break. There were however some problems with the management of emergency calls, mostly of a procedural nature (Benuska, 1990).

3.3 Available Standards and Remedial Measures

3.3.1 Standards, Codes, and Specifications

As previously noted, communication lines and equipment are not regulated by local building codes. However, the structures used, mostly central offices, are regulated by local building codes. While these codes provide adequate protection against collapse and loss of life, they do not necessary insure that equipment is adequately protected. In particular, lateral drift limitations may not be adequate for the purpose, and special site hazards should receive consideration. Thus, in some areas, local codes should be supplemented with more severe stiffness requirements, and some requirements for consideration of site related risks (i.e. liquefaction and proximity to active faults). This may increase building cost, but since the building cost is only a small portion of the total plant cost (Faynsod, 1987), the impact on total cost would be very moderate. Adequate seismic design of central office buildings could possibly be accomplished via assignment of appropriate importance factors (seismic hazard exposure) to these buildings.

18

There are industry standards for equipment and tiedown. For example, Bell Communications Research (Bellcore) Technical Reference "Network Equipment Building System (NEBS)" (Bellcore, 1988) is a generic equipment standard which contains provisions for earthquake resistant design. The standard is updated on an annual basis and contains design as well as testing requirements. The standard requires conservative consideration of amplification by the building, as well as the equipment. Figure 2 shows the recommended design spectra for four earthquake zones. The standard calls for equipment qualification by vibration tests and contains a list of test facilities available in the United States. Another Bellcore document entitled "National Security Emergency Preparedness-NSEP", which was prepared for internal use by the Bell Operating Companies, deals with the operational aspects of the emergency response of the communication network to earthquakes.

The Bellcore technical references are recommendations which are available to the various divested regional Bell Operating Companies and to other telephone companies and may be used at the company's discretion. They are being used for new construction as well as for retrofit of old facilities by Pacific Bell, and similar provisions are used by other telephone companies throughout California. However the recommendations are not universally used throughout the United States, and there are many central offices in other U.S. regions which are vulnerable to earthquakes. Pacific Bell also requires equipment supplies to comply with the Bellcore recommendations. There is no evidence that standards similar to the Bellcore recommendations are being utilized by companies other than the divested regional Bell operating companies.

3.3.2 Remedial Measures

Available protective measures are standardized bracing systems developed by Bellcore and others, vibration-resistant equipment meeting Bellcore or similar requirements, standby power supplies in the form of batteries and generators, and slack in cables and ducts crossing active faults and at the connections to buildings and other fixed structures. In addition there is considerable redundancy in communication systems. Inoperative portions of the system are automatically bypassed, so that the impact is minimized. Additional safeguards

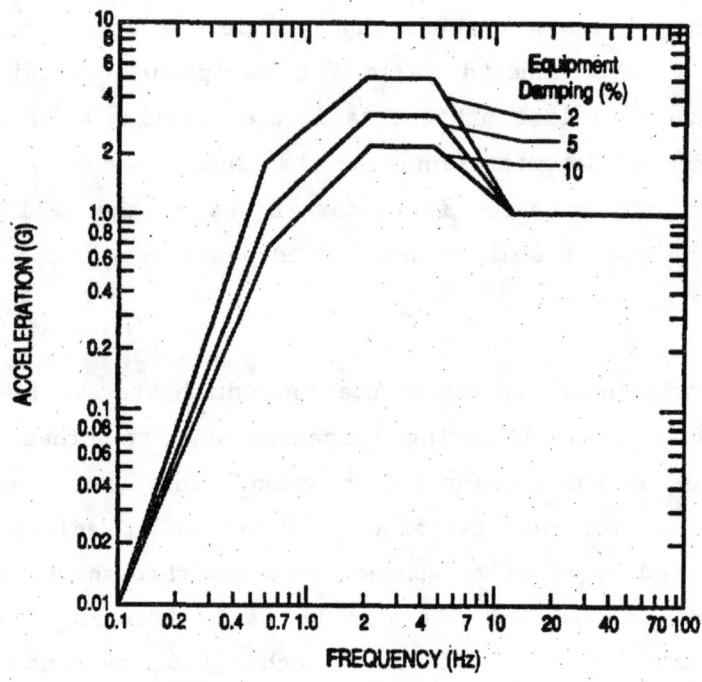

Installation Location	Zone			
	4	3	2	1
Upper Floors	1.0	0.6	0.4	0.3
Midlevel Floors	0.9	0.5	0.3	0.2
Ground-Level and Basement Floors	0.8	0.4	0.2	0.1

Figure 2: Seismic Design Response Spectra Recommended for Telecommunication Equipment (reproduced from Bellcore, 1988)

are emergency operating procedures which will block incoming non-emergency calls and place other restrictions on telecommunication uses in impacted areas.

Decentralization of the system after divestiture may have complicated the situation in terms of installation of earthquake-resistant equipment and retrofit of vulnerable installations. The equipment which is now supplied from various domestic and foreign sources on a competitive basis may also complicate the quality control problem. On the other hand there is more redundancy as a result of de-centralization, and adequate standards are available for the purchase of earthquake resistant equipment. Competition among sources of supply also has benefits for regional companies which developed good product specifications to meet their particular need, because they can select the products which best meet their particular needs, rather than modifying generic products for their purpose.

While there is decentralization and redundancy as a result of divestiture, there is also an opposing trend toward concentration because of the great carrying capacity of fiber-optic cables. This trend could potentially increase the seismic vulnerability of some central offices.

4. PROTECTION OF FEDERALLY-CONTROLLED SYSTEMS

4.1 Introduction

The information in this section was supplied by the Federal Agencies listed. No independent evaluation by NIST was attempted.

This report identifies three types of lifelines which could be required to meet Federally imposed standards: (1) lifelines which are owned and operated by the Federal Government, such as the Western Area Power Administration electrical transmission lines; (2) lifelines which are owned and operated by others, but have a major contractual obligation to serve the Federal Government, such as the Southeastern Power Administration electrical transmission and the Federal Telecommunication System; and (3) lifelines which are constructed with Federal funds, but not owned by the Federal Government, and which do not have a major contractual obligation to serve the Federal Government, such as electrical transmission lines financed by the Rural Electrification Administration.

Only a small portion of the power transmission facilities actually utilized by the Federal Government fall in one of the categories listed above. Thus it is important to realize that even if all the facilities over which the Federal Government can exercise some control were adequately protected against seismic risk, a large portion of the systems on which the Federal Government depends may still be vulnerable to earthquake damage.

4.2 Hazard-mitigation Measures Implemented by Federal Agencies

DEPARTMENT OF ENERGY

General mitigation criteria:

The Department of Energy has a policy of overall risk assessment. Earthquakes are one of the risks. As a minimum, all construction has to comply with the Uniform Building Code. For high-hazard situations (nuclear, dams, etc.) a ground spectrum for the site is developed for the maximum credible earthquake.

Electrical transmission and microwave towers are not generally designed for seismic forces; however, in individual cases, these forces would be considered in the risk assessment (for instance for secondary risks such as conflagration of fuel storage facilities). There would be retrofitting if deemed necessary. The power administrations under DOE (Western, Alaska, Southeastern) set their own earthquake hazard-mitigation policies.

Western Area Power Administration:

There are no special standards for electrical transmission lines in seismic areas. However, transmission and microwave towers are generally not built in the immediate vicinity of a fault. Electrical substations are designed to withstand seismic forces if this is warranted (as in California). Generally there is built-in redundancy in the electrical transmission and communication

22

systems. Equipment for electric power substations in areas of high seismic risk (Tracey & Redding, CA) must meet the requirements of IEEE 693-1984 [10], using an electrical equipment response spectrum for up to 0.5 G maximum high-frequency ground acceleration (zero period response acceleration in IEEE terminology). In central United States areas a less severe response spectrum is used (0.2 G). Connections for mounting are specified accordingly by the suppliers. Buildings are designed in accordance with UBC, which is also used to determine seismic zoning. No retrofit of existing structures is planned.

Alaska Power Administration:

The Alaska Power Administration subcontracts all the construction to the Corps of Engineers in Anchorage, who in turn is using seismic design as appropriate. In the great 1964 earthquake there was no major problem with power transmission and only one tower was lost.

Earthquake resistant design is used in accordance with Corps of Engineers design criteria, using the services of a geologist and a geotechnical engineer in the more vulnerable areas. Seismic zoning follows USGS recommendations. There are no plans for retrofit, but seismic design has been used since 1964.

Some of the lines and all communications are owned and operated by private utility companies (Chugak Electric and RCA). There are no major transmission lines in permafrost. Most communication systems are via satellite.

Southeastern Power Administration

The Administration does not own transmission or communication facilities. Power transmission is sub-contracted to private utility companies. No attempt is made to impose seismic design requirements on these private utility companies.

BUREAU OF RECLAMATION

There is a general policy of using maximum ground accelerations of 0.15 G in moderate risk areas and 0.2 G in high risk areas for electrical equipment.

23

There is no definite policy on buildings housing the equipment. Discussions are now under way on seismic design criteria for the buildings.

There is presently no retrofit program. Most power transmission and communications facilities, which used to be operated by the Bureau of Reclamation, were transferred to DOE.

TENNESSEE VALLEY AUTHORITY

Transmission lines and towers, including microwave towers, are not specifically designed to resist earthquake forces. However, if a line were to traverse an area of known liquefaction hazards, pile foundations would be used.

Transformers and other large equipment are clamped down against a 0.2 G seismic acceleration (installations are open-air). The western portion of the TVA area is considered vulnerable. Not much is done to tie down other equipment. There is standby power by redundant electrical power supplies but there are no standby generators. There have been no seismic events during the Authority's 60-year history.

CORPS OF ENGINEERS

Seismic provisions are used in the design of dams and power plants. There are no special provisions for transmission towers. However, this does not mean they are not designed for seismic forces. It is generally the practice to follow local codes when the peak ground accelerations are thought to exceed 0.15G.

Criteria used for equipment tie down are given in references Army TM 5-809-10, 1982, Army TM 5-809-10-1, 1986, U.S. Army Corps of Engineers, CEGS 15200, 1985 There is presently no retrofit program.

NAVAL FACILITIES COMMAND (NAVFAC)

Generally, attention is being given to seismic design in new construction, but no retrofitting is being done.

Standards TM 5-809-10, 1982, and TM 5-809-10-1, 1986, deal with seismic standards for buildings. Military Handbooks 1004/4 and 1004/2, 1987, for electrical utilization systems and power distribution systems, mention adverse weather condition but not earthquakes. NAVFAC P-355.1, 1985, provides design criteria for seismic support for existing electrical-mechanical equipment and utilities.

RURAL ELECTRIFICATION ADMINISTRATION

The Rural Electrification Administration finances, but does not own, electrical systems. Transmission lines are generally supported by wooden poles which are not vulnerable to earthquakes. No retrofit is contemplated or could be legally enforced. Compliance with IEEE standards is not required.

Bulletin 65-1, "Rural Substations", 1978, covers earthquake-resistant design for buildings in seismic areas and recognizes earthquakes as a problem. However, IEEE 693-1984, which would insure adequate design of substation equipment, is not mentioned.

U.S. FOREST SERVICE

The Forest Service uses microwave towers which are not very tall. To the extent that these are designed by the Forest Service, UBC is used which includes seismic design provisions. To the extent that their design and construction is contracted out, this is not necessarily the case. No retrofit of facilities is contemplated.

GENERAL SERVICE ADMINISTRATION (GSA)

GSA has 10-year FTS 2000 contracts with AT&T and U.S. SPRINT for voice, data, and video sevices. These services are provided through a system of digital switches and fiber-optic cables and have a high degree of redundancy and survivability. However, the contracts do not include explicit requirements for earthquake-resistant design of facilities.

GSA thinks that there is sufficient redundancy in the system to work around areas affected by earthquakes and sees no need to spend the extra funds that would be associated with special requirements for earthquake protection.

U.S. POSTAL SERVICE CORPORATION

The Postal Service uses commercial power supply and common carriers for most communications. However, it has a telephone communication system. One example of communication centers in areas of high seismic risk is the center in San Mateo, CA. The building housing the center was built in 1985, and it is reasonable to assume that it was built in compliance with local building codes which provide adequate seismic protection. The equipment was purchased from suppliers in accordance with GSA schedules, and no special seismic requirements were stipulated at the time of its purchase. There has been no study to determine whether the facility is vulnerable to earthquake damage.

5. SUMMARY AND RECOMMENDATIONS

5.1 System Vulnerability

Electrical Transmission Systems:

Records from past earthquakes indicate that electrical transmission lines are not very vulnerable to earthquake damage. However, earthquake damage occurred in areas of unstable soils. Substations were damaged in many earthquakes as a result of breakage of porcelain components, inadequate tiedown, gasket leaks, inadequate slack in electrical lines, and inadequate clearance between system components. Distribution lines are not very vulnerable, however some damage occurred as a result of tangled wires and toppled platform-mounted transformers.

Telecommunication Systems:

Past earthquake records indicated the central telecommunication offices and the equipment they house are in some instances vulnerable to earthquake damage. A large number of central telecommunication offices in the United States are located in seismically active areas. Telecommunication lines and microwave towers did not suffer much damage, except in areas of unstable soils.

Regional Aspects of Vulnerability:

On the West Coast where there is a high frequency of seismic events, the need for earthquake resistant construction is recognized and electrical transmission and telecommunication systems are generally designed and built to be earthquake resistant. In seismically vulnerable areas on the East Coast and in the Central U.S. this is frequently not the case. The vulnerability of these systems is further increased by the fact that historical records of major earthquakes in the Eastern and Central U.S. indicates that, because of their greater focal depth and seismotectonic conditions which differ from those on the West Coast, these earthquakes affect larger areas than those on the West Coast. Thus some of the system redundancies which tend to protect electrical transmission and telecommunication lifelines on the West Coast may not function in a major earthquake in the Eastern or Central U.S.

5.2 Existing Guidelines, Standards, and Protective Measures

Electrical Transmission Systems

ANSI/IEEE Standard 693-1984 is a recommended practice for seismic design of substations. However, West Coast power companies consider this standard inadequate and use their own specifications. Required performance of substation components can be specified using provisions such as those presently used on the West Coast together with a specified maximum ground acceleration. There are also criteria in the model building codes (UBC, 1988, BOCA, 1987, Southern, 1989), the SEAOC, 1985, Tentative Lateral Force Requirements, and the NEHRP, 1989, Recommended Provisions which could be used to design foundation tiedowns and components of electrical substations, however the provisions in these documents are primarily used in conjunction with building codes which do not necessarily apply to the design of electrical substations. Several design manuals published by military agencies also contain provision that are applicable to electrical substations. Utility companies on the West Coast replace the ANSI/IEEE standard with their own specifications to insure adequate earthquake resistance.

Buildings housing equipment can be designed in accordance with local building codes or other provisions for the design of earthquake resistant buildings. Technology is available, and presently used in new construction on the West Coast, which can provide adequate components, housing and tiedown for substation equipment. Adequate clearances and slack in connecting lines can also be provided, even though some configurations which provide protection against earthquake loads are not always the most efficient from the standpoint of electrical performance. However, present technology does not provide adequate protection of, or a satisfactory substitute for, porcelain components. Thus even substations which are designed in accordance with the best available technology are susceptible to some earthquake damage. Efforts are now under way to develop materials which can replace presently used porcelain.

Telecommunication Systems

Standards for telecommunication centers and generic technical requirements for new equipment are being developed by Bellcore and some other industry organizations. Standards developed for central office equipment and its tiedown and bracing systems are designed to provide adequate protection. These standards are being updated on an ongoing basis. Bellcore also developed operational procedures for earthquake emergencies and hazard mitigation. Compliance with these industry standards is voluntary. The standards are followed closely on the West Coast, where existing facilities are also often retrofitted to comply with the standards. In other parts of the country the emphasis on earthquake protection is generally less and varies among regions. Central offices are presently designed in compliance with local building codes which call for earthquake resistant design. However, some experts recommend that stiffness requirements for these buildings exceed local building code requirements. This recommendation could possibly be implemented via assignment of appropriate importance factors (seismic hazard exposure) to these buildings.

There is presently no indication that facilities designed in accordance with Bellcore or similar recommendations, using adequate design ground accelerations, would be vulnerable to earthquake damage. However, it is important to note that as a result of divestiture, equipment is now being purchased from a variety of U.S. and foreign sources. Unless seismic requirements are incorporated in the specifications for this equipment, it may be vulnerable to earthquake damage. There is a trend toward centralization resulting from the introduction of high-capacity fiber-optic cable which could increase the risk associated with the failure of some central offices.

5.3 Federal Practices

There are three types of lifelines which could be required to meet Federally-imposed standards: lifelines which are owned and operated by the Federal Government; lifelines which are owned and operated by others, but have a major contractual obligation to serve the Federal Government; and lifelines which are constructed with Federal funds but not owned by the Federal Government and which do not have a major contractual obligation to serve the Federal Government.

Most new construction projects for electrical transmission systems which are owned and operated by the Federal Government are designed to provide some degree of earthquake resistance. However, their design and construction specifications are generally not as stringent as those followed by private utilities on the West coast. Also, different agencies follow different rules. At the present time not much retrofitting of inadequate **existing** facilities is planned.

There is no evidence that Federal Agencies require compliance with the Bellcore recommendations when telecommunication equipment for seismically vulnerable areas is purchased, or that G.S.A. schedules for the purchase of this equipment incorporate earthquake-resistant components.

There is also no evidence that privately owned and operated electrical transmission and telecommunication facilities which have a major contractual obligation to the Federal Government are required to be protected against earthquake hazards. However, some of the operators of these facilities may provide such protection on their own initiative.

Electrical transmission facilities constructed with Federal funds and to Federal specifications, but not serving the Federal Government, were required to provide some earthquake protection; but the scope of these requirements is limited and does not include references to some of the standards and practices discussed in this report.

5.4 Recommendations

1. While standards and guidelines are available for electrical transmission systems, they are generally not all-inclusive and need to be updated. To have an inclusive up-to-date standard several different standards should be updated and combined.   It is not clear whether this is also necessary for telecommunication systems.

It is recommended that inclusive Federal guidelines for recommended seismic performance requirements for electrical transmission and telecommunication systems be developed under the auspices of the Interagency Committee on Seismic Safety in Construction. Since many Federal Agencies are opposed to a Federally imposed uniform standard, it is recommended that the guideline consist of two parts: (1) performance criteria; and (2) a model standard document.

The performance criteria should specify required levels of performance, without specifically stating how that performance is to be achieved. This will permit various agencies to make their own choice whether to achieve the required performance by following a specific standard or by developing their own provisions. The model standard is envisioned as a compilation of existing standards or specifications, complemented where necessary by additional provisions. An agency could follow the performance criteria and use their own standards, or alternatively could follow the model standard.

As a first step in the development of this document, joint Federal-industry workshops should be organized in order to assess the adequacy of existing standards and guidelines and identify the areas that need more work.

6. ACKNOWLEDGEMENT

Review comments and suggestions by James Cooper, ICSSC, John Foss, Bellcore, Jeremy Isenberg, Weidlinger Associates, Robert Kassawara, EPRI, Shih-Chi Liu, NSF, Anshel Shiff, Stanford University, Masanobu Shinozuka, Princeton University, Ken Sullivan, FEMA, Alex Tang, ASCE Technical Council, and Larry Wong, Pacific Bell, are gratefully acknowledged.

6. REFERENCES

ANSI A.58.1-1982, "Minimum Design Loads for Buildings and Other Structures", American National Standards Institute, 1982.

ANSI/IEEE Standard 693-1984, "IEEE Recommended Practices for Seismic Design of Substations", Institute of Electrical and Electronic Engineers, August 1984.

ANSI/IEEE Standard 344-1987, "IEEE Recommended Practices for Seismic Qualification of Class 1E Equipment for Nuclear Power Generating Stations", Institute of Electrical and Electronic Engineers, 1987.

Army TM 5-809-10, NAVFAC P-355, Airforce AFM 883, Chap. 13, "Seismic Design for Buildings", February 1982.

Army TM 5-809-10-1, NAVFAC P-3555.1, AFM 88-3, Chap. 13, "Seismic Design Guidelines for Essential Building", February, 1986.

ASCE Committee on Dynamic Analysis, "The Effect of Earthquakes on Power and Industrial Facilities and Implications for Nuclear Plant Design", American Society of Civil Engineers, 1987.

ASCE Technical Council on Lifeline Earthquake Engineering, "Advisory Notes on Lifeline Earthquake Engineering", American Society of Civil Engineers, 1983.

Bellcore, "Network Equipment-Building System (NEBS) Generic Equipment Requirement", Technical Reference TR-EOP-000063, Morristown, NJ, March 1988.

Benuska, 1., Editor,"Loma Prieta Earthquake Reconnaissance Report", Earthquake Spectra, Supplement to Vol.6, El Cerrito, CA, May, 1990.

Building Seismic Safety Council, "Abatement of Seismic Hazards to Lifelines: An Action Plan", Federal Emergency Management Agency FEMA 142, Earthquake Hazards Reduction Series 32, August 1987.

Building Seismic Safety Council, "NEHRP Recommended Provisions for the Development of Seismic Regulations for New Buildings", Federal Emergency Management Agency, FEMA 95, Earthquake Hazards Reduction Series 17, February 1986 (amended 1988).

Building Officials & Code Administrators International, Inc., "the BOCA Building Code/1987", 10Th Edition, Country Club Hills, IL, December, 1986.

City of Los Angeles, Department of Water and Power, "Seismic Design Criteria for Equipment", August 1988.

EPRI (EQE Inc. & Bechtel Power Corporation), "The 1986 North Palm Spring Earthquake: Effect on Power Facilities", Report EPRI NP-5607, Electrical Power Research Institute, January 1988.

EPRI (EQE Inc.), "Performance of Industrial Facilities in the Mexican Earthquake of September 19, 1985", EPRI Report NP-4605, Electrical Power Research Institute, June 1986.

EPRI (EQE Inc.), "Investigation of the San Salvador Earthquake of October 10, 1986, Effects on Power and Industrial Facilities", Report EPRI NP-5616, Electrical Power Research Institute, February 1988.

EPRI (EQE Inc.), "Effects of the 1985 Mexico Earthquake on Power and Industrial Facilities", Report EPRI NP-5784, Electrical Power Research Institute, April 1988.

EPRI (EQE Inc.), "Reconnaissance Investigation of the March 2, 1987, New Zealand Earthquake", Report EPRI NP-5970, Electrical Power Research Institute, August 1988.

Faynsod, L., "Available Criteria, Methods, and Techniques for the Design and Construction of New Seismic-Resistant Lifelines", FEMA 137, Earthquake Hazards Reduction Series 28, July 1987.

Foss, J.W., "Earthquakes and Communications Lifelines", National Earthquake Engineering Conference, Knoxville, TN, September 15, 1981.

Filled Polymer Electrical Insulator, US Patent No. 4210774.

High Strength Alumina Ceramic Product, US Patent No. 4183760.

International Conference of Building Officials, "Uniform Building Code", 1988 Edition, ICBO, Whittier, CA, May 1,1988,

Isenberg, J., "Seismic Performance of Telecommunications Equipment", Tsukuba Science City, December, 1984.

Mirzad, S.S., "Survey of Ongoing Activities in the Abatement of Seismic Hazards to Communications Systems", Abatement of Seismic Hazards to Lifelines, FEMA 137, Earthquake Hazard Reduction Series 28, July 1987.

Military Handbook, "Electrical Utilization Systems", MIL-HDBK-1004/4, October 1987.

Naval Facilities Command, "Seismic Evaluation of Supports for Existing Electrical-Mechanical Equipment and Utilities", NAVFAC P-355.1, March 1975.

Rural Electrification Administration, "Design Guide for Rural Substations", REA Bulletin 65-1, June 1978.

Seismology Committee, Structural Engineers Association of California, "Tentative Lateral Force Requirements", SEAOC, Sacramento, CA, October 1985.
Shinozuka, M., "Scientific and Engineering Information Needs in Abatement of Seismic Hazards to Telecommunication Systems", FEMA 137, Earthquake Hazards Reduction Series 28, July 1987.

Shiff, A.J., Earthquake Effect on Electric Power Systems, <u>J. Power Div.,ASCE</u> V. 99, No. PO2, Nov., 1973.

Southern Building Code Congress International, "Standard Building Code", 1988 Edition (with 1989 revisions), SBCCI, Inc., Birmingham, AL, January,1989,

Steinhardt, Otto, W., "On-Going Activities in Abatement of Seismic Hazards in Electric Power Systems", FEMA 138, Earthquake Hazards Reduction Series 29, July 1987.

Stone, W.C., Yokel, F.Y., Celebi, M., Hanks, T., and Leyendecker, E.V., "Engineering Aspects of the September 19, 1985 Mexico earthquake", NBS Building Science Series 165, May 1987.

Tang, A., "Research and Development in Seismic Mitigation of Telecommunication Systems in Canada", Proc. US-Japan Workshop on Seismic Behavior of Buried Pipelines and Telecommunications systems", Tsukuba City, Japan, December 1984.

U.S. Army Corps of Engineers, CEGS-15200, "Guide Specification, Military Construction"..., Seismic Protection for Mechanical, Electrical Equipment, September 1985.

Yanev, P.I., ed., "Earthquake Engineering Research Institute Reconnaissance Report, Miyagi-Ken Oki, Japan, December 1984.

Appendix

DISCUSSIONS WITH STAFF MEMBERS FROM FEDERAL AND OTHER AGENCIES

Michael E. McCafferty, Western Area Power Administration

Electrical Transmission: Generally there are no special standards for seismic areas, but structures are generally not built in the immediate vicinity of a fault. However electrical substations are designed to withstand seismic forces if this is warranted (in California). No special design provisions are applied in the case of microwave towers, except that they are not constructed in the vicinity of a fault. Generally there is built-in redundancy in the electrical transmission and communication systems.

Terry Burley (also Don Warner), Western Area Power Administration

Equipment for electric power substations in areas of high seismic risk (Tracey & Redding, CA) must meet the requirements of IEEE 693-1984, using an electrical equipment response spectrum for up to 0.5 G maximum ground acceleration. In central US areas a less severe response spectrum is used (0.2 G). Connections for mounting are specified accordingly by the suppliers. Buildings are designed in accordance with UBC, which is also used to determine seismic zoning. No retrofit of existing structures is planned.

E.C. Pritchett, U.S. Army Corps of Engineers

While seismic provisions are used in the design of dams and power plants, the respondent is not aware of specific provisions for power transmission and communication facilities. Applicable standards can be scanned for relevant provisions.

Ivar Paavola, U.S. Army Corps of Engineers, Structures Branch

There are no special provisions for transmission towers. However, this does not mean they are not designed for seismic forces. It is generally the practice to follow local codes when the peak ground acceleration is expected to be less than 0.15 G, and to employ a geologist and a geotechnical engineer to develop a ground spectrum when the peak ground acceleration is thought to be higher. There are design criteria for equipment tiedown.

John Tyrrell, Naval Facilities Command

Generally attention is being given to seismic design in new construction but, no retrofitting is done.

James R. Hill, Department of Energy

DOE has a policy of overall risk assessment. Earthquakes are one of the risks. As a minimum, all construction has to comply with UBC. For high-hazard situations a ground spectrum for the site is developed, looking at time frames as long as 10,000 years. Transmission towers are not generally designed for seismic forces, however in individual cases these forces would be considered in the risk assessment (for instance for secondary risks such as conflagration of fuel storage facilities).

Lee A. Julson, Bureau of Reclamation

There is a general policy of using maximum ground accelerations of 0.15 G in moderate risk areas and 0.2 G in seismic areas for electrical equipment. Policies on buildings housing the equipment are now under discussion. There is no retrofit program. Most power transmission and communication facilities which used to be operated by the Bureau of Reclamation were transferred to DOE.

Lee A. Belfore, Rural Electrification Administration

The Rural Electrification Administration finances, but does not own, electrical systems. Transmission lines are generally wooden poles which are not vulnerable to earthquakes. However Bulletin 65-1 [33] recommends earthquake-resistant design in seismic areas. No retrofit is contemplated or could be legally enforced. Compliance with IEEE standards not required.

James R. Talbot, Soil Conservation Service

The Soil Conservation Service is not responsible for any power or communications line.

James Lloyd, Southeastern Power Administration

Southeastern Power does not own transmission or communication facilities. Power transmission is sub-contracted to private utility companies. No attempt is made to impose seismic design requirements on these companies.

Richard Lee, George R. Hanks, Tennessee Valley Administration

Richard Lee: There is no seismic design for transmission lines and towers, including microwave towers. However, if a line were to traverse an area of known liquefaction hazards, pile foundations would be used. There is a requirement to tie down transformers.

George Hanks: transformers and other large equipment are clamped down against a 0.2 G seismic acceleration. The western portion of the TVA area is considered vulnerable. Not much is done to tie down other equipment. There is standby power by redundant electrical power supplies but there are no standby generators. There were no seismic events during the Authority's 60-year history.

Don Russell, U.S. Army Corps of Engineers District Office, Anchorage, AK.

Earthquake resistant design is used in accordance with Corps design criteria, using a geologist and geotechnical engineer in the more vulnerable areas. Seismic zoning follows USGS recommendations. There is no retrofit, but seismic design has been used since 1964. Some of the liens and all communications are by private utility companies (Chugak Electric and RCA). There are no major transmission lines in permafrost. Most communication systems are via satellite.

Clyde Shumway, U.S. Forest Service

The Forest Service uses microwave towers, not very tall. To the extent that theses are designed by the Forest Service, UBC is used which includes seismic design provisions. To the extent that their design and construction is contracted out, this is not necessarily the case. No retrofit is contemplated.

William Cunnane, General Services Administration

GSA has private telecommunication lines with 58 main nodes and awards 30-day tariffs for circuits. While some facilities are hardened for emergency use, private industry tariffs do not include explicit requirements for earthquake resistant design of facilities. GSA thinks that there is sufficient redundancy in the system to work around areas affected by earthquakes and sees no need to spend the extra funds that would be associated with special requirements for earthquake protection.

Updated information provided on 9.5.'90 by Bruce Brignull, FTS 2000 Manager, GSA:

GSA has 10-year FTS 2000 contracts with AT&T and U.S. SPRINT for voice, data, and video sevices. These services are provided through a system of digital switches and fiber-optic cables and have a high degree of redundancy and survivability.

Ronald Schwitz, U.S. Postal Service

The postal service uses commercial power supply. However, it has a telephone communication system. One of the centers is in San Mateo, CA, an area of high seismic risk. The building housing the center was built in 1985, and it is reasonable to assume that it was built in compliance with local building codes which provide adequate seismic protection. The equipment was purchased from suppliers in accordance with GSA schedules and no special seismic requirements were stipulated at the time of its purchase. There was no study to determine whether the facility is vulnerable to earthquake damage.

John W. Foss, Bellcore

Bell Communications concentrate research on intra-state systems, AT&T on interstate systems. Generic standards are provided in Technical Ref. TR-EOP 000063, Issue 3 "Network Equipment Building System" March '88 [32] (document updated on ongoing basis). The document contains generic-type requirements, which are referenced when equipment acquisitions are made (now some of the equipment is manufactures overseas). Emergency earthquake-response procedures are now being prepared and will be incorporated in the document entitled "Bellcore/BCC National Security Emergency Preparedness (NSEP) Operational Guidelines". Generally, seismic design is implemented on the West Coast, but it is not clear whether in other area seismic design has priority enough to be implemented. Generally competitive pressures are not as important as PUC approval when it comes to expenditures on seismic safety. Microwave towers are not a major concern, however problems may arise with guyed towers (unbalanced cable pull) and microwave towers on top of buildings.

John Hinton, Bellcore

A operational guideline on earthquake emergency procedures and hazard mitigation was prepared in May 1989. It focuses on operational procedures, such as network management techniques. Other standards are promulgated in the form of "information letters" issued by Bellcore. These are accepted by all the regional companies created after the breakup of AT&T. They are not necessarily implemented by competing companies. Several trade associations may be interested in standards: The Association of Federal Communication Engineers, USTA (United States Telephone Association), NATA (National Telephone Companies Association).

Larry Wong, Pacific Bell

The environmental Standards in Ref. [32] are imposed by Pacific Bell. All equipment installed is now shake table tested to very exacting spectra. All buildings meet local codes, but they are heavier, and therefore stiffer than conventional. Battery racks and tiedowns have to meet the same standards. Slack is provided in trunk lines and loops. Equipment is now purchased from various U.S. and foreign sources. Pacific Bell shares information on earthquake resistant design and acquisition of qualified equipment with other companies inside and outside California.

Bob Kassawara, Electrical Power Research INstitute

EPRI recently conducted a series of studies to evaluate the performance of electrical equipment in recent earthquakes. The study reports concentrate on generating facilities, but they contain a fair amount of information on substations and switchyards. They are:

NP 5970	New Zealand Earthquake
NP 5784	Mexico Earthquake
NP 5607	Palm Springs Earthquake
NP 5616	El Salvador Earthquake
NP 4605	Mexico Earthquake

EPRI also started to work on specific projects related to switchyard seismic ruggedness, but these will take several years because priority is given to nuclear power projects in the EPRI seismic program.

Anshel Shiff, Stanford University

Problems in the Midwest and East are potentially larger than those on the West Coast, because larger areas could be affected by earthquakes. If good design provisions were implemented, normal equipment attrition would eventually result in earthquake resistant installations even without retrofit. Even after the Sylmar experience, when equipment was properly tied down, an "adequately designed" modern substation suffered major damage in the Palm Spring earthquake, much of it due to porcelain failures.

In communications, the fact that Bell no longer is sole equipment producer results in much more vulnerable equipment as there is an incentive to cut corners.

Luis Escalante, L.A. Department of Water and Power

There is an "Power and Water Earthquake Engineering Forum" of a California Water and Power working group and an EPRI steering committee on seismic ruggedness of substations. There will be (was) a workshop for California utilities organized by Bob Kassawara in October 1988. The various interested groups are communicating and using similar specifications.

The Department uses its own specifications in preference to prevailing standards which are not considered adequate. The Devars substation was constructed using lessons from San Fernando (Sylmar substation). Yet the switchyard suffered severe damage in the Palm Springs earthquake. The problem is porcelain and there is as yet no good answer to it. There is a conflict between seismic safety and electrical efficiency. Frequently electrical engineers do not like clearances and geometries worked out for seismic protection. These conflicts have to be resolved. Hammering and forces exerted on connecting lines are major problems and the solutions are not necessarily cut and dry.